Making Biodiesel

How to brew your own fuel at your backyard

Samuel Bell & Alan Delfin

this document is not allowed unless with written permission from the publisher. All rights reserved.

The information provided herein is stated to be truthful and consistent, in that any liability, in terms of inattention or otherwise, by any usage or abuse of any policies, processes, or directions contained within is the solitary and utter responsibility of the recipient reader. Under no circumstances will any legal responsibility or blame be held against the publisher for any reparation, damages, or monetary loss due to the information herein, either directly or indirectly.

Respective authors own all copyrights not held by the publisher.

The information herein is offered for informational purposes solely, and is universal as so. The presentation of the information is without contract or any type of guarantee assurance.

The trademarks that are used are without any consent, and the publication of the trademark is without permission or backing by the trademark owner. All trademarks and brands within this book are for clarifying purposes only and are the owned by the owners themselves, not affiliated with this document.

Contents

Disclaimer and Terms of Uses

The Author and Publisher has strived to be as accurate and complete as possible in the creation of this book, notwithstanding the fact that he does not warrant or represent at no time that the contents within are accurate due to the rapidly changing nature of the Internet. While all attempts have been made to verify information provided in this publication, the AuthorandPublisher assumes no responsibility for errors, omissions, or contrary interpretation of the subject matter herein.

Acknowledgement

I have taken effort in this project. However, it would not have been possible without the kind support and help of many individuals who contribute to this project, especially my mother, who played a key role to write this book...

INTRODUCTION TO BIODIESEL

Biodiesel is a clean fuel that consists of several raw materials, such as used vegetable oils, virgin vegetables, animal fats, and yellow fats. What distinguishes biodiesel from the use of pure vegetable oil is a process called transesterification, a process that will be discussed later in this publication.

Biodiesel is one of the biofuels that can be used to meet the energy needs of society. For example, ethanol is a very common biofuel produced from raw sugar cane, potato, corn, and manioc. But biodiesel is not ethanol because ethanol is a renewable biofuel for gasoline engines.

The interest in biodiesel increased due to rising fuel prices, the need for energy independence (biodiesel can be generated from domestic sources) and the need for environmentally friendly energy sources. ,

Advantages and disadvantages of biodiesel

The use of biodiesel as a fuel source has many benefits. Firstly, biodiesel is cleaning diesel engines from deposits accumulated over time. Secondly, biodiesel can be cheaper than traditional diesel fuel because raw materials such as used vegetable oil can be quickly and cheaply available in places like restaurants. However, the time and energy spent on creating biodiesel can make it more expensive than the available diesel fuel. Third, biodiesel is a clean, renewable source of combustion energy. Fourth, low-volume biodiesel production does not require more power than output. According to Biodiesel.Org, the leading organization in research and marketing of biodiesel:

Biodiesel has one of the most substantial "energy balances" among all liquid fuels. We get 4.5 units of energy for each group of fossil energy needed for the production of biodiesel. This takes into account sowing, harvesting, fuel production and fuel transportation for the end user.

Part of this calculation is probably because raw materials, such as used vegetable oil, can be obtained nationally, which means that no oil transportation from other countries is required, which reduces the cost of energy and fossil fuels. Also, biodiesel can be used alone in modified diesel engines, which means that no oil can be obtained, usually purchased in other countries.

However, biodiesel has some disadvantages. The disadvantage is that small biodiesel from a biodiesel producer requires a lot of time and energy. Another lack of biodiesel is that it produces less energy than traditional fuel sources such as diesel and gasoline. Additionally, if waste vegetable oil is used as a raw material, for example, large quantities of water have probably been used for vegetable cultivation, which perhaps had adverse effects on the environment in terms of swelling and pesticides. Also, washing fuel requires significant water consumption.

In any case, more than ever, it is essential to understand the different fuel sources available for vehicles. Therefore, we must understand and know which options we have as consumers and members of the biosphere to make responsible decisions

Sustainability

• The National Biodiesel Council has prioritized sustainability in the areas of mitigation of climate change, human rights, food security and respect for all natural resources. That is why the industry has adopted and follows the guiding principles that demonstrate our commitment to a whole series of sustainable development principles.

• Biodiesel producers already provide a very sustainable fuel, and these principles are another way to ensure that our sector continues to improve quality of life, protect the environment and strengthen the environment as it develops savings.

• Biodiesel improves air quality; it is renewable and creates a green collar of jobs in our communities. NBB is committed to maintaining biodiesel at the forefront of sustainability.

Energy balance

• The energy balance of biodiesel is very high: a study recently published by the University of Idaho and the United States. The Ministry of Agriculture shows that for each unit of fossil energy needed for the production of biodiesel, the return is 5.54 units of energy. Soybean oil-based biodiesel has a high energy balance since solar energy is the primary source of energy for the cultivation of soybean.

• "Energy Balance" takes into account planting, harvesting, fuel production and fuel transportation to the end user. Thanks to modern techniques of breeding and energy efficiency, the energy balance of biodiesel continues to improve.

• On the other hand, biodiesel based on conventional fossil fuels has a negative energy balance.

Water storage

• Crops are not sprayed or planted exclusively for the production of biodiesel. The conversion of these co-products and by-products uses very little water - the entire US In 2008, the biodiesel industry consumed less water than was needed to irrigate two golf courses in the United States.

• 1998 Jointly Produced American Edition The end-to-end analysis of the Ministry of Agriculture and Energy for the production of biodiesel has shown that it reduces waste water by 79% and the output of 96% hazardous waste about oil diesel.

Preserving the Earth

• USDA reports that the cultivated US crop production area has not increased since 1959.

• In the United States, no significant changes are expected in land use that would endanger environmentally sensitive land due to biofuels. There are necessary federal and state laws that help ensure that these countries remain unchanged.

• Crop production in the United States. UU. It has a significant tendency to use multiple conservation practices, and progress in agriculture leads to higher yields and lower inputs with the same area.

• The Food and Agriculture Organization of the United Nations (FAO) has calculated that only 3,700 million hectares of 10,400 million hectares are used in agriculture, and only 1% of this area is used for biofuels, which includes and ethanol.

Security of food supply

• Biodiesel is not produced by mincing soybean in fuel. Soy has two components: fatty meal and protein. Food is the majority of soybean, and it is used in food and feed for livestock. Biodiesel uses only a portion of soybean oil.

• By creating a new market for co-produced soybean oil, the total pea value increases, and part of the food becomes more competitive in terms of costs for protein markets. This has a positive net impact on food supply.

• Biodiesel does not affect food prices as large food companies would like to believe. For example, in the last quarter of 2008, production of biodiesel accounted for nearly 60 million gallons of water per month, but in the same period, soybean products were sold at the lowest levels. And if this is not enough, even when commodity prices are falling, food prices have barely moved.

• Biodiesel produced from American soybeans uses only about 3 percent of soybean oil each year.

• Biodiesel uses only a portion of soybean oil, leaving all the available protein to feed cattle and humans.

• In 2008, biodiesel produced from soybean produced enough soybean meal for the equivalent of 115 billion meals of hungry developing countries in developing countries.

Diversity

• Biodiesel is the most diversified fuel on the planet. Produced from renewable sources available in the region that abounds in the United States, Including soybean oil, other vegetable oils, recycled fats, bovine fats, and other fats.

• Increased demand for biodiesel encourages research and investment in the development of new biodiesel production materials, such as algae, camelina, jatropha, other dry land crops, and wastes such as trap fat. The result is that we will see additional quantities of raw materials from the soil to the hills or low production and using innovative technologies.

According to a study conducted by the National Renewable Energy Laboratory in Golden, Colorado, the domestic raw materials used for biodiesel were 1.6 trillion gallons (including fats, animal fats, and vegetable oils). NREL predicts that the natural growth and expansion of existing raw materials (soybean, oilseed rape, and sunflower) will increase the supply of raw materials to an additional 1.8 billion gallons by 2016.

Cleaner effects on air and health

• DOE and USDA say biodiesel reduces carbon dioxide and greenhouse gas emissions by 78 percent. Biodiesel also significantly reduces emissions regulated by the Environmental Protection Agency and has a direct impact on human health.

• Biodiesel is the only alternative fuel to voluntary EPA Level I and II testing to quantify emissions and health effects.

• Respiration of particles is a danger to human health. Emissions of biodiesel particles from the exhaust are about 47% lower than total diesel particulate emissions.

• Biodiesel emissions show a dramatic decrease in the level of polycyclic aromatic hydrocarbons (75 to 85%) and polycyclic aromatic nitric hydrocarbons (90% at the trace level), identified as potentially carcinogenic compounds.

• Because of the health benefits of biodiesel, some chapters of the American Lung Association have committed to supporting the use of alternative fuels2.

Will I save money making my biodiesel?

Many suppliers offer biodiesel processors in the British market. Many of them can be used without little technical knowledge about the process. It is often mentioned that the production of biodiesel could cost less than 20 ppm per liter, which, depending on its annual mileage, could be seen in black in less than a year.If your annual mileage is too high or you are considering the production of biodiesel for your small fleet, keep in mind that since January 2010, each person has been given an allowable amount of 2,500 liters a year to produce biodiesel. You are not required to notify HMRC unless your production level exceeds the previous quota, but if you exceed this limit, you will have to pay the customs duty on the fuel you produce.

Where can I buy biodiesel?

Biodiesel is currently available in most countries that produce oilseeds, and many farmers use biodiesel as a means to stimulate production and increase public awareness. However, farmers' approach to biodiesel for use on farms is another dimension that requires reflection and encourages the potential of sad irony. Farmers who cultivate biodiesel crops in isolated rural areas may not have access to fuel. Unless farmers intend to produce biodiesel on site, or at a more extensive local biodiesel production facility, many farmers may not be able to use the fuel on which they are working. On its website, the National Biodiesel Board lists about 1,400 suppliers in the United States, but some have stopped distributing biodiesel. Almost all countries have at least one pumping station that offers a mixture of biodiesel, although not necessarily at a practical distance from most potential customers. The website publishes a map of biodiesel stores throughout the country. The committee recommends that regional fuel distributors be asked for more biodiesel delivered locally (from NBB, 2009).

Uses of biodiesel

1. **Hydrogen production for fuel cell vehicles.**

It was a beautiful story of the month: InnovaTeka researchers have developed hand-sized microreactors capable of converting biodiesel (or any other liquid fuel) into a flow of hydrogen that can be used in a nearby fuel cell. Chevron has already invested $ 500,000 in developing fuel cell vehicle service stations. InnovaTek hopes to eventually integrate microreactors into vehicles, which will fill cars with biodiesel while offering more efficient and even cleaner electrical units. See the full story here. This is a potential goal of InnovaTek: to integrate its technology with automobiles, where renewable energy sources can be transformed into a movement, avoiding combustion and the production of exhaust in its entirety, and driving an engine. Much more efficient (Imagine for a moment filling the biodiesel and that the electric motor is perfectly fused). InnovaTek plans to obtain a commercial license for microreactors for 2009.

The weight of less than half a kilogram, a square piece of shiny steel (top), contains a series of microchannels containing patented catalytic sites. Each micro tube helps convert (or reform) the continuous hydrogen streams of fuels such as gasoline, diesel, vegetable oil, biodiesel, propane, natural gas or by-products — glycerol from the production of biodiesel.

Although hydrogen produces an apparatus considered to be "the energy of the future," it faces significant development problems. Hydrogen is not an important energy carrier. It has a relatively low energy density, is difficult and dangerous to transport and it is difficult to find storage in vehicles that run on hydrogen (the first F-cell of Mercedes had a range of 110 miles). The fuel fueling infrastructure does not exist either.

More importantly, we have not yet implemented a renewable energy source for the production of hydrogen.But that did not stop us from building a hydrogen fuel cell car. GM, Ford, Honda, Hyundai, and Toyota have prototypes, and Mercedes is at the end of 2007.Given all this, the Innovatek reactor could revolutionize the country's energy and transport infrastructure.

Innovatek has already signed a $ 500 million joint development agreement with Chevron to apply hydrogen fuel refining technology. (If you think its high, in September 2006, the Navy also awarded Innovatek a $ 1.8 million contract for the development of portable cargo systems that marines usually carry on foot.) One of Innovateka's board members commented on its ability to reduce the cost of hydrogen generation: "A lower system size, a reduced catalyst volume and a more efficient process by InnovaTek technology is another important step in making the hydrogen economy to science becomes a commercial reality, "he said.

While the InnovTtek reactor can work with several sources of non-renewable hydrocarbons, both they and the potentially revolutionary Coskata Biofuels are explicitly interested in sustainable energy, even to the point that they prefer biodiesel in their tests. Innovatek also said that only biodiesel is more effective: it contains fewer impurities and reforms at lower temperatures than petrodiesel.

Let me beat the substitutes here: we will not in any way supply all US cars to biodiesel, even if we use this kind of technology. I am also interested in exploring what products derived from micro reactor products and how they will be collected and used. But without the ability to print biodiesel algae or other important productive raw materials as possible solutions, and based on the inherent cold of this device, I think we can all be cautiously optimistic.

2. **Clean up the oil spill.**

Biodiesel is known to be ecologically benign, but who would have thought that this could help purify oil spills? Biodiesel was tested as a potential cleaning agent for contaminated crude oil and was found to increase the recovery of crude oil from artificial sand pools (i.e., beaches). It is also used in commercial bio-solvents that are effective in thickening crude oil and allow the surface of the water to be cleaned. In 1997, the Cytosol product was licensed by the California Department of Fish and Wildlife as a cleaning agent.

3. **Electricity generation.**

In addition to the production of hydrogen fuel for vehicles (see point 1), fuel cells have applications that produce energy that can be used by biodiesel. The military has already invested $ 1.8 million in mobile power generation using this technology and could be available for civilian applications shortly.

Biodiesel is already used in conventional power generation. In 2001, UC Riverside installed a 6-megawatt backup generator system containing 100% biodiesel. The project was successful, and the typical work smoke from diesel generators was virtually nonexistent. Biodiesel can be used in safety systems where the reduction of essential emissions is significant: hospitals, schools and other facilities commonly found in residential areas. It can also be used in addition to solar energy in off-grid homes (for more information, see Kemp 2006).

Petroleum electricity was used in 2006 for 115,370,000 barrels of oil, a quantity that could be replaced entirely by biodiesel production in the United States.

4. **Warm up at home**

Bio heat had become increasingly popular in recent years, and biodiesel can be used as home heating oil in domestic and commercial boilers (No. 2 fuel oil is almost identical to petrodiesel). While 20% of the biodiesel blend (B20) can be used without modification, a higher mix can affect rubber seals and older appliance seals. Blends with high biodiesel content will also clean the fuel lines, which may improve heating efficiency, but may initially cause fouling of the fuel filter. A mixture of 20% biodiesel will reduce emissions of sulfur dioxide (SO_2, acid rain) and nitrogen oxides (NO_x, pollutants contributing to ozone at the soil level) by 20% in all air configurations.

There may be a company in your area that specializes in bio-heating. See Portland Green Heat to see an example.

5. **Camping: cooking and lighting**.

It is possible to use biodiesel instead of kerosene in some lamps and some non-stove furnaces. For example, the BriteLyt Petromax fuel cell battery will work well with biodiesel (almost everything will burn). BriteLyt also produces more fuel furnaces. But at 4 pounds, it's not something you want to use with a backpack.

I've always wondered if the traditional furnace can process biodiesel. For example, MSR Whisper Lite International and Primus Multifuel are designed to work on many fuels, including gasoline, diesel, and kerosene.

BIODIESEL PRODUCTION

The process of producing biodiesel involves a chemical reaction. This means that the biodiesel industry is a chemical industry. Those involved in the production of biodiesel must have a good knowledge of the chemical structure they have, to ensure that the products produce quality fuel safely. Biodiesel is an alternative fuel for diesel engines obtained by the chemical reaction of vegetable oil or animal fat with an alcohol such as methanol or ethanol. In words, the result is

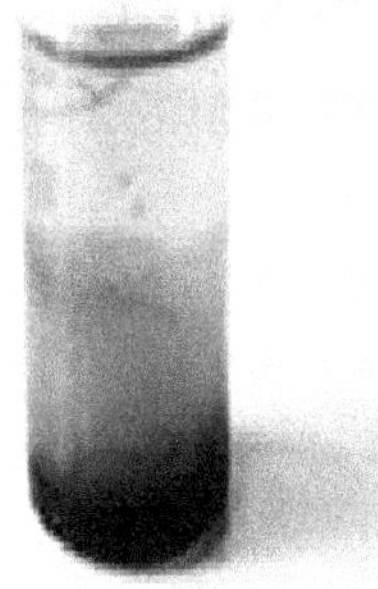

Oil + alcohol → biodiesel + glycerin

The image shows a bottle of biodiesel and glycerol (also called glycerol).

Bio diesel is a layer of lighter color on the top. The darker glycerol oil is sunk to the bottom.

It is essential to understand that unmodified vegetable oil, sometimes called vegetable oil (SVO) or waste vegetable oil (WVO), is not biodiesel. Some people have used SVO or WVO in diesel engines with varying degrees of success. The main problem is the high viscosity and low volatility of unmodified vegetable oils. Manufacturers of US UU engines they recommended the use of SVO and WVO. Biodiesel is generally preferred over SVO and WVO because the chemical reaction turns oil or fat into compounds that are closer to hydrocarbons in regular diesel.

A chemical reaction that converts vegetable oil or animal fat into biodiesel is called "transesterification." It is a long-term process of combining a chemical compound called "ester" and alcohol to obtain another ester and another alcohol. Oils and fats are included in the family of esters. When they react with methanol or ethanol, they produce methyl or ethyl esters and new alcohol called glycerol or, more often, glycerin.

Vegetable oils and animal fats used to produce biodiesel can come from almost any source. All of these products consist of chemicals called triglycerides so that biodiesel can be obtained from soybean oil, canola oil, beef and fat, and even exotic oils such as oil of nuts avocado oil.

Even cooking oils or waste oils can be used for biodiesel production. However, these oils pose particular problems for the production of biodiesel because they contain pollutants such as water, meat residues and mating that must be filtered before the oil becomes biodiesel.

Methanol is the most common alcohol used to produce biodiesel. Sometimes it is also called methyl alcohol or wood alcohol. It is very toxic, and ingestion of a single spoon can cause blindness or even death. Hazardous exposure can also occur when inhaling methanol or methanol through skin contact. In the United States, ethanol usually costs more than methanol and is less used in biodiesel production. It is alcohol that is found in alcoholic beverages, so it is not toxic in small quantities. However, it is subject to stringent state regulations because of the tax requirements for alcoholic beverages.

A chemical reaction used to produce biodiesel requires a catalyst. The catalyst is generally a chemical added to the reaction mixture to accelerate the reaction. As the catalyst is not consumed in the reaction, it will remain somehow in the end. In biodiesel production, the actual compound that catalyzes the reaction is called methoxide. The usual method of producing methoxide is the dissolution of sodium hydroxide or potassium hydroxide in methanol. Major manufacturers buy a methanol solution of sodium in methanol with which it is much safer to work.

High-quality biodiesel is defined by the D6751 of the American Society for Testing and Materials (ASTM). Fuel testing to check compliance can be expensive, especially for small producers, but this is the most reliable way to ensure that fuel consumers have access to high-quality fuel.

In addition to reducing costs, another definite advantage of inedible oils for biodiesel production is that food is not consumed for fuel production. These reasons, among others, have led to examining medium and large-scale biodiesel production in several countries, using insect oils such as castor oil, Tung, cotton, jojoba, and jatropha. Animal fats are also an attractive option, especially in countries with substantial livestock resources, although prior treatment is necessary because they are strong. Also, highly acidic fats from cattle, swine, poultry, and fish can be used.

Microalgae seem to be a significant alternative to future biodiesel production due to high oil yields; however, it must be taken into account that only certain types are useful for the production of biofuels.

Although the properties of the oils and fats used as raw materials can be distinguished, the properties of biodiesel must be identical, by the requirements established by international standards.

Raw materials for the production of biodiesel

Raw materials for the production of biodiesel are vegetable oils, animal fats and short chains of alcohol. The most used oils in the world's biodiesel production are oilseed rape (mainly in EU countries), soya (Argentina and the United States), palm trees (Asia and Central America) and sunflower, although other oils, including peanuts, are also used. , flax seeds, safflowers, used vegetable oils, as well as animal fats. Methanol is the most commonly used alcohol, although ethanol can also be used. Since the cost of production and commercialization of biodiesel (mainly due to oil prices) is the primary concern, the use of rancid vegetable oils has been explored with good results for several years.

Typical oilseeds useful for the production of biodiesel

Canola and canola oil

Canola adapts well to low fertility soil but with a high sulfur content. With a high oil yield (40-50%), it can be grown as a winter cover crop, allowing for double cropping and seeding.

It is the essential raw material for biodiesel production in the European Community. However, in some Central and South American countries, technical restrictions have been imposed on planting and harvesting, mainly because of lack of adequate information on fertilization, handling, and storage of seeds (seeds). Are very small and require specialized agricultural equipment). Also, low prices compared to wheat (its main competitor for harvest) and low production per unit area limited its use.

Rapeseed oil has a high nutritional value compared to soybean; it is used as a protein supplement in animal cultures.Sometimes, canola and rapeseed are considered synonymous; rape

Canadian Low Acid Oil is the result of the genetic modification of canola in the last 40 years in Canada to reduce the erucic acid and glucosinolate content in oilseed rape, which is uncomfortable for both animal and human consumption.

Canadian Low Acid Oil is the result of the genetic modification of canola in the last 40 years in Canada to reduce the erucic acid and glucosinolate content of oilseed rape, which causes discomfort in human and animal consumption.

Canola oil is highly valued for its high quality. With olive oil, it is considered one of the best foods for cooking because it helps to lower cholesterol levels in the blood.

Raw materials for biodiesel production

Soy

It is a legume native to East Asia. Depending on the environmental conditions and the genetic species, the plants have significant differences in height. The main soybean producing countries are the United States, Brazil, Argentina, China, and India.

Soybean biodiesel production, in addition to glycerin, also provides other useful by-products: flour and soy pellets (used as livestock feed) and flour (which contains high levels of lecithin and protein). The grain yield varies between 2000 and 4000 kg/ha. The seeds were very rich in protein; the oil content is about 18%.

Palm oil

Palm oil is a tropical plant reaching a height of 20 to 25 m with a life cycle of about 25 years. The complete production is carried out eight years after planting.

From the fruit, two types of oil are obtained: palm oil, pulp, and palm oil, from nuts (after extraction of the oil, cakes of palm seeds released as food for the livestock). Several types of high oil yield have been developed. Indonesia and Malaysia are the leading manufacturers.

International demand for palm oil has steadily increased in recent years. The oil has been used for cooking, as a raw material for the production of margarine and as a supplement to butter and baked goods.

It is important to note that pure palm oil is semi-solid at room temperature.

(20-22_C), and in many applications, it is mixed with other vegetable oils, sometimes partially hydrogenated.

Sunflower

The sunflower "seedlings" are a fruit, a whole wall (shell) that surrounds the seed in the grain.

The great importance of sunflower lies in the excellent quality of the edible oil extracted from its seeds. It is very popular from nutritional quality, taste, and taste. Also, after extraction of the oil, the remaining cake is used as feed for livestock. It should be noted that sunflower oil has shallow linoleic acid content and can, therefore, be stored for a long time.

Sunflower adapts well to adverse environmental conditions, does not require specialized agricultural equipment and can be used for corn seeds and seeds. The yield of liquid hybrids is between 48 and 52%.

Peanuts

Groundnut quality is strongly influenced by weather conditions during harvest.

Groundnuts are mainly used for human consumption, in the production of peanut butter and as an ingredient for sweets and other processed foods. For the production of oil, poor quality peanuts are used (including waste from the confectionery industry), whose demand is stable on the international market — peanut oil used in cooking blends and as a flavor enhancer in the confectionery industry.

The flour residue, after extraction of the oil, is of high quality and contains a lot of protein. In the form of pellets, is used as feed for livestock.

Flax

Flax is a temperate climate plant with blue flowers. Litter is made from the stem of the plant, and the seed oil is called flaxseed oil, which is used in the production of colors. Flaxseed has nutritional value for human consumption because the source of polyunsaturated fatty acids is necessary for human health. Also, the remaining cake, after extraction of the oil, is used as feed for livestock.

The plant adapts well to a wide range of temperature and humidity; However, high temperatures and heavy rains do not favor high yields of seeds and fiber.

Flaxseed contains between 30 and 48% of the oil and the protein content between 20 and 30%. It is important to note that linseed oil is rich in polyunsaturated fatty acids and that linoleic acid accounts for 40-68% of the total amount.

Saffron

Saffron is well adapted to dry conditions. Although the grain yield per hectare is low, the oil content of the seed is high, 30 to 40%. It, therefore, has economic potential for drylands. Currently, saffron is used in the production of oils and flour and the diet of birds.

There are two species, one rich in monounsaturated fatty acids (oleic acid) and the other with a high percentage of polyunsaturated fatty acids (linoleic acid). Both varieties are low in saturated fatty acids. Saffron oil is of high quality and little in cholesterol.

In addition to being used for human consumption, it is used in the production of paints and other coatings, varnishes and soaps.

It is important to note that safflower oil is secreted by hydraulic presses, without the use of solvents, and refined by conventional methods, without antioxidant additives. Saffron flour is high in fiber and contains about 24% protein. It has been published as a protein supplement for livestock feed

Castor Seeds

The castor oil plant grows in tropical climates, with temperatures between 20 and 30 ° C; He cannot stand the cold. It is important to note that when the seed starts to germinate, the temperature should not be lower than 12 ° C. Plants need a warm and humid period during the vegetative phase and a dry season for maturation and the harvest. It takes a lot of suns and adapts well to several types of soil.

The total amount of precipitation during the growing cycle should be between 700 and 1400 mm; although it is drought resistant, castor oil needs at least five months of rain per year.

Castor oil is a triglyceride whose main ingredient is ricinoleic acid (about 90%). The oil is an insect and toxic because of the presence of 1-5% ricin, an atoxic protein that can be eliminated by cold compression and filtration. The presence of hydroxyl groups in its molecules makes it unusually polar compared to other vegetable oils.

Tung

Tung is a tree that adapts well to tropical and subtropical climates. The optimum temperature for the Tung is between 18 and 26 ° C, with low annual rainfall.

During the harvest season, dry nuts fall from tungsten and accumulate on the ground. Nut production begins three years after planting. Tung nut oil is inedible and used in the manufacture of paints and varnishes, especially for marine use.

Cotton

Among non-food products, cotton is the best-selling product. It is produced in more than 80 countries and is distributed worldwide. After harvest, you can exchange raw cotton, fiber or seeds. In cotton mills, the grains and seeds are separated from the natural cotton. Cotton fiber is processed for the manufacture of fabrics and spikes, for use in the textile industry. Also, cottonseed oil and flour are obtained from the seed; the latter is rich in protein and is used in livestock feed and after processing for human consumption.

Jojoba

Although jojoba can survive extreme drought, irrigation is necessary to achieve an economically viable yield.

Jojoba needs a warm climate, but severe episodes are required for flower blooming. Rain must be deficient during the harvest season (in summer). The plant reaches its full productivity ten years after planting. Jojoba oil is mainly used in the cosmetics industry. So his market is quickly saturated.

Jatropha

Jatropha is a shrub that adapts well to dry environments. Jatropha curcas is the most famous variety. Requires little water or extra care; therefore, it is suitable for areas of low fertility. Productivity can be reduced by erratic rainfall or high winds during the blooming season. Yield depends on climate, soil, precipitation, and treatment during sowing and harvesting. Jatropha plants become productive after 3 or 4 years and have a lifespan of about 50 years.

The oil yield depends on the extraction method; which is 28 to 32% by pressure and up to 52% by solvent extraction. Since the seeds are toxic, jatropha oil is impossible. The toxicity is a consequence of the presence of curcasin (globulin) and jatrophic acid (toxic like ricin).

Avocado

The avocado is a tree 5 to 15 m high. The weight of the fruit is between 120 and 2.5 kg and the harvest period varies from 5 to 15 months. The avocado fruits ripen after the harvest, not on the tree. The oil can be obtained from fruit pulp and cores. It has a high nutritional value because it contains essential fatty acids, minerals, proteins, and vitamins A, B6, C, D and E. The content of saturated fatty acids in the fruit pulp and the oil is low. On the contrary, it is very rich in monounsaturated fatty acids (about 96% is oleic acid). The oil content of the fruit varies from 12 to 30%.

Microalgae

Microalgae have a high potential for biodiesel production, as the oil yield (in liters per hectare) could be 1-2 times higher than the return of other feedstocks. The oil content is generally between 20 and 50%, although it may exceed 70% in some species. However, it is important to note that not all microalgae are suitable for biodiesel production.

High levels of CO_2, water, light, nutrients, and minerals are essential for microalgae growth. Production processes take place in fish ponds and photobiological reactors.

Collection of raw material

Used vegetable oil

The first step is to find oil. Display the map of your area and select a quadrant on one side of the city.

Take a road map to go down all the main streets where the consumer stores are. Prepare a plan by traveling the shortest distance possible while traveling all significant roads with companies that may have a grease container. The types of companies that can contain containers are fast-food chains, pizzerias, family restaurants, cinemas, bowling alleys, hotels, playgrounds, bars, bars with ready meals, schools, hospitals, donuts, department stores, and even large industrial producers. Potatoes. There are countless others!

When planning a trip, you need to create a sample collection box and fill it with empty sample containers.

A sample collection box

Locate one or two large tables and place a box with a plastic garbage bag to hold the water. You want to lubricate the oil on the entire surface of the box so that the plastic bag is tight on the sides.

A nice plastic box would be ideal, but I use a card to throw it away instead of cleaning it. Then you will need a bunch of small oil tankers with closing caps. Containers are safe in the dishwasher and bite nicely on the lid. You can use any tank that will allow you to collect a tiny spoon of full oil. What is a full spoon? Well, you only need 1 ml to perform a titration test of the oil, but usually, I take about 200 ml if I want to show other tests.

By using containers that are all identical, the operation becomes better organized, and the containers can be stacked together, requiring less space.

Fill up as many collection containers as you think you need once or twice. I have about 30 sample containers, but I usually only need 15 to 20 places to go hunting.

You should also carry a roller coaster and a Flowmaster with a permanent marker as a blade. Whenever you take a sample of oil, the container must be marked to determine where it is obtained.

You may want to start keeping records; it will save you a lot of time and allow you to run later.

Execute the planned trip

Now that you're all ready to go, we'll discuss some of the details about the physical attributes of the sample collection process.

Consider driving conditions. Many food institutions are geographically clustered. This means that you will enter and exit these facilities every few minutes and that in the immediate vicinity, traffic can have nightmares. I usually collect samples late in the evening to avoid traffic jams. The task of three or four hours quickly decreases to less than an hour if the streets are empty. My region is relatively safe from a crime standpoint, which should be taken into account.

Sort the good from the bad

When you return with all your samples, perform a standard titration for each one and record this information.

Select the oil samples you think are the best and find the phone numbers for each location.

 Now it's time to make phone calls and ask permission to start taking oil. Many people start having problems. Some people are great people and prefer to go to the restaurant, while others become very nervous. There are those who like to go to lunch and talk to the manager; some send letters and others like me make a simple phone call.

The way you will get a license will depend a lot on the region in which you operate and your personal preferences.

Overall, I need to visit about ten restaurants to find about 40 gallons of high-quality oil a week. I managed to get permission to take this oil about nine times out of 10 with a two-minute phone conversation.

Here are some tips on what to say and what not to say.

• Be very direct. "My name is Joe, and I would like to allow myself to take some of your used vegetable oil."

• If / when asked why you want, give brief and sincere answers. Make sure you include something good for global warming and oil independence.

• If you discover uncertainties in their reactions, be sure to tell them that you never spill a drop.

• Be nice if you get refused.

Transporting

Disposal of waste oil from containers behind the restaurant and transfer of this oil to the processor may be returned to work, and there is not much information available on how this is done. The amount of fuel you need to move and your geographic position will play a significant role in choosing the technique.

Whatever your decision to carry your used oil, you should seriously consider the following questions.

1. Professionalism goes a long way. Being courteous, presentable and keeping your word to the restaurant

manager can make or break the source of the oil in an instant.

2. Cleanliness is a top priority in almost every restaurant. Always carry a cleaning kit with you if you drop or spill oil on the floor. As the container is closer to the traveler's trip, the cleaner will usually be, and you will need to clean it better.

3. Secure the load! Always make sure the oil you are carrying is in your vehicle. Each driver can be the victim of accidents and close conversations, and a break, even at low speed, can cause a significant change in the physical position of the oil tank (s). All tanks or pumps must be glued, fastened or attached in such a way as to

4. Avoid spills during a quick maneuver or accident.

For those who only need 20 to 30 gallons a week to meet their fuel needs, one of the most economical ways to collect oil is to ask the restaurant to return the oil in its packaging. But some clean your fryers while they are hot and have to use a metal container.

If you only need a small volume of oil every week, this is a perfect way.

• Individual containers do not require investment

• Reasonably secure because of a low amount

• Not practical for larger quantities of over 30 or 40 gallons.

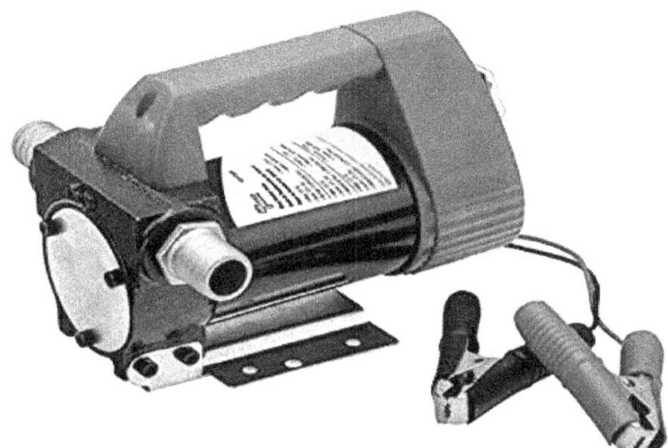

12-volt electric oil pumps are another option. There are different models of these oil pumps, but my research has shown that they are not suitable for the collection of waste vegetable oil. One of the reasons is that they are designed to change the engine oil where the pump works only for a short time. The intermittent pump cannot tolerate the continuous nature of oil collection, and as a result, the engine usually heats up and goes bad.

For small amounts of transmission, one model that we have heard is that high-quality products are Fill-Rite. These pumps are made of cast iron and come with some pretty robust and high-quality engines. While I do not own one of these pumps, I own a fuel filler pump and can guarantee the high-quality standards that enter these units.

• A nice compact unit

• It is not right in cold climates when the oil is thicker

• Uses 12 volts

• Suitable for low to medium volume

Garbage pumps work well for oils that are liquid in warm climates but begin to lose the suction rate at lower temperatures when the oil is thicker. They can also be quite expensive.

• Garbage pumps are standard and easy to find.

• They are loud and attract unwanted attention.

• Most gas and diesel engines require extremely expensive.

• Do not work in colder climates.

• They are complicated and require maintenance.

• It can easily emulsify the oil.

I believe that the method of collecting the vacuum is far superior to any other technique that is less than a dedicated and manufactured truck that is adapted for this purpose.

 In the purest sense, the device is called Super Sucker and does precisely what its name implies. It was made by pulling a deep vacuum on the container, then channeling the space through the hose and valve. The operator dives the tube into the waste oil tank and then opens the valve.

The vacuum in the tank will attract dense oil. Super Sucker does not differ much from the usual wet and dry vacuum. The most significant difference is that it is much stronger and uses the positive displacement pump instead of the blower to extract the deep vacuum needed to collect waste oil in colder climates. Nobody sells Super Sucker, as far as I know, so if you want to go this way, you will have to build it.

Though it seems complicated, it's nothing more than a small 12-volt vacuum pump connected to the air compressor tank with a vacuum gauge.

• Easy to build

• Cheap if you can find a pressure vessel for holding a deep vacuum

• Reliable because it has several moving parts

• Cold, dense oil will suck

• Works for all requirements for a quantity of five gallons to five thousand gallons

• It will not emulsify oil such as a centrifugal pump

• Quiet work. Does not increase noise

<u>**Health and safety**</u>

Proper clothing for the process includes closed-toed shoes, goggles, aprons, and gloves. Also, make sure all pump is in perfect working order when collecting mostly used oils.

First Aid Measures:

- Inhalation:

Remove from area to fresh air. Seek medical attention if symptoms persist.

- Eye Contact:

Flush eyes with a heavy stream of water for 15 to 20 minutes. Seek medical care if symptoms persist or worsen.

- Skin:

Wash contaminated areas of the body with soap and water.

- Ingestion:

Give one to two glasses of water to drink. If gastrointestinal symptoms develop, seek medical attention. NEVER GIVE ANYTHING BY MOUTH TO AN UNCONSCIOUS PERSON!

BIODIESEL CHEMISTRY

A typical biodiesel molecule looks like the structure below. It is mainly a long chain of carbon atoms, with atoms attached to hydrogen, and at one end what we call an ester functional group (shown in blue).

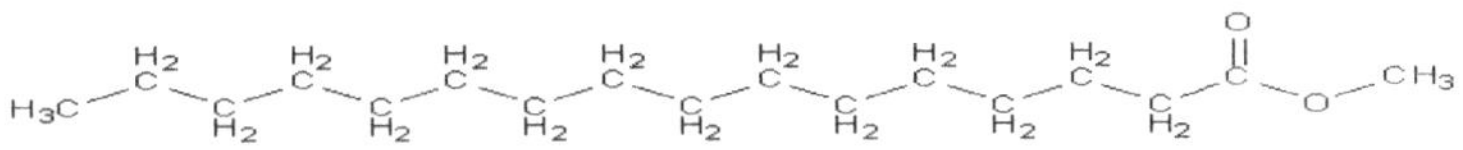

Diesel engines can burn biodiesel without modifications (except to replace some rubber tubes that can be softened with biodiesel). This is possible because biodiesel is chemically very similar to the conventional diesel fuel shown below. Note that regular diesel has a long chain of carbon and hydrogen atoms, but there is no group of esters in blue.

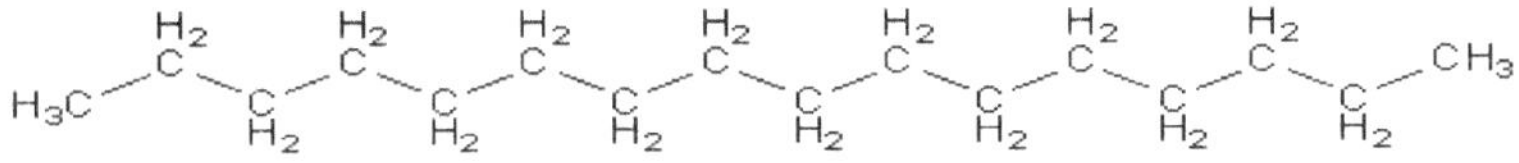

The first diesel engines did not work with "diesel" fuel, but with vegetable oil, whose sample molecule is shown below. Note that it also has long rows of carbon and hydrogen atoms, but it is about three times larger than normal diesel molecules. It also has functional groups of esters (in blue), such as biodiesel.

This more significant amount of vegetable oil means that in cold climates, it gels, which makes its use in the engine difficult. By converting it into biodiesel, it turns it into smaller molecules, closer to the size of ordinary diesel, so it has to cool before the vegetable oil before it starts to gel.

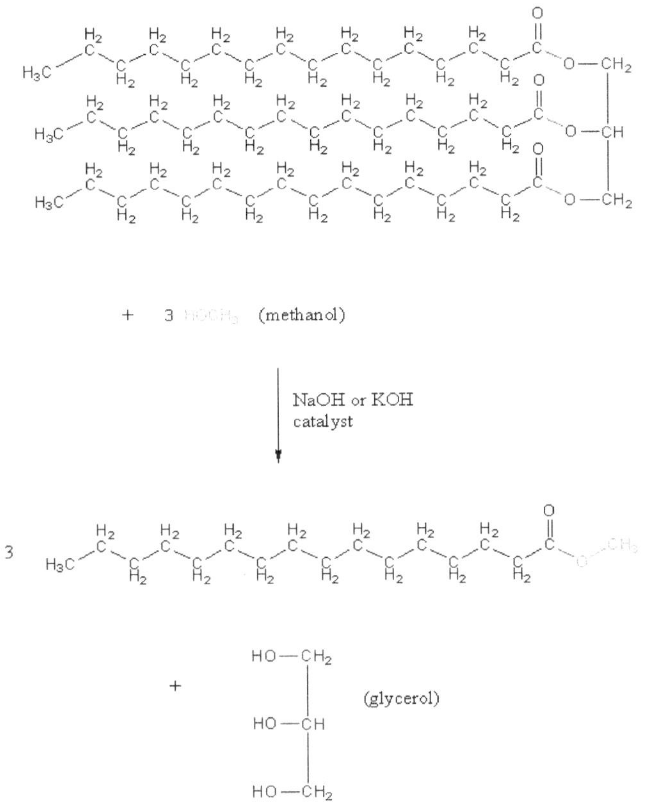

+   3 HOCH₃ (methanol)

NaOH or KOH
catalyst

3

HO—CH₂
+   (glycerol)
HO—CH

HO—CH₂

Chemical conversion of vegetable oil into biodiesel.

Vegetable oil, like biodiesel, belongs to the category of compounds called esters. Therefore, the conversion of vegetable oil into biodiesel is called a transesterification reaction. For this reaction, you need to use methanol (shown in green), which causes

55

the red bonds to break in the structure below. This interrupts the blue section, like the backbone of the molecule, which turns into glycerol. The red bonds that entered the glycerol backbone were linked to methoxy groups, which are shown in green in the final structure, which comes from methanol:

Organic Chemistry

• Acid: a corrosive substance that increases the concentration of hydrogen ions (H_3O +) in water. Its pH is less than 7.

• Base: a caustic substance that increases the level of hydroxide ions (OH-) in water. pH is higher than 7.

• Catalyst: a material that facilitates the reaction between other substances by creating an alternative route for the appearance of the reaction.

• Alcohol - A substance that contains hydroxyl compounds (-ROH).

• Ester - acid related to alcohol

• Esterification - ester formation. Example: The acid joins the alcohol and forms an ester.

• Transesterification: the transformation of one type of ester into a different kind of ester

• Triglycerides: the main component of fat consists of three molecules of fatty acids in combination with the glycerol alcohol molecule (Figure 1). Triglycerides form a large part of many types of lipids (fats).

• Fatty acids are straight-chain carboxylic acids (saturated or unsaturated). They are obtained from the hydrolysis of a fat or can be synthesized from two carbon units (acetyl or malonyl-CoA) in the liver, the milk gland and, to some extent, the fatty tissue. Almost everyone has an equal number of carbon atoms. Individual fatty acids, free fatty acids (FFA) or non-esterified fatty acids (NEFA) circulate mainly about albumin. They are essential metabolic fuels.

DIESEL ENGINE

The diesel engine also known as a compression ignition engine is an internal combustion engine that uses compression heat to ignite to fuel combustion that is injected into the combustion chamber. This is in contrast to a spark ignition engine, such as a petrol engine (gasoline engine) or gasoline engine (using gas fuel instead of gasoline), which uses a spark plug to ignite the blend. Air and fuel. Rudolf Diesel developed the engine in 1893. The diesel engine has higher thermal efficiency than any other internal or external, internal combustion engine because of the high degree of compression. Low-speed diesel engines used on ships and other applications where the total engine weight is relatively small may have a thermal efficiency of more than 50%. The diesel engine is manufactured in two-stroke and four-strokeversions.

They were initially used for more efficient replacement of stationary steam engines. Since 1910 they have been used in submarines and ships. Use in locomotives, trucks, heavy machinery and power plants continued later. The 1930s began to be applied slowly in some cars. Since the 1970s, the use of diesel engines in large vehicles, both on and off the road. Today, exhaustion of fossil fuel reserves, increased demand for diesel and uncertainty as to their availability is a source of global concern. From a car standpoint, reducing diesel fuel consumption means improving engine performance to reduce energy losses during combustion. In the diesel engine, the combustion and emission characteristics are influenced by the fuel spraying 2, the nozzle geometry, the injection pressure, the shape of the inlet opening and other factors. To improve the mixing of fuel and air, it is essential to understand the fuel spraying and the process of forming the spray. To date, to improve combustion and particle emissions, many researchers have studied the characteristics of spray behavior through experimental and theoretical approaches. Atomization is a process of a chemical reaction between the fuel injected and the compressed air in the combustion chamber, where the injection of high-pressure fuel decomposes into a mist or a very fine droplet. This process is essential for producing fuel ready for combustion. The smallest drop comes out of the nozzle, evaporates more quickly and ignites when injected into the engine cylinder. An efficient atomization process will reduce HC production and emissions. Many factors influence the atomization process, such as the injection pressure, the temperature in the combustion chamber, the geometry of the piston surface and the geometry of the mouthpiece of the nozzle.

Main tpes diesel engine

There are three primary size groups of diesel engines based on power: small, medium and large. Small engines have an output power of fewer than 188 kilowatts or 252 horsepower. This is the type of diesel engine most commonly produced. These engines are used in cars, light trucks and specific agricultural and construction applications, such as small fixed electric generators (such as pleasure boats) and as mechanical drives. These are usually direct injection engines, aligned on four or six cylinders. Many are turbochargers with a subsequent cooler.

Medium-sized engines have a power of 188 to 750 kilowatts or 252 to 1,006 horsepower. Most of these engines are used in heavy trucks. In general, it is a direct injection, linear, six-cylinder engine, turbocharged and post-cooled. Some V-8 and V-12 engines are also part of this size group.

Large diesel engines have a power of more than 750 kilowatts. These different motors are used for applications in the ship, locomotive and mechanical drives, as well as for power generation. In most cases, these are direct injection systems, turbochargers, and subsequent cooling. They can operate at only 500 rpm when reliability and durability are essential.

Two-stroke and four-stroke engines.

Diesel engines are designed to operate in a two- or four-stroke cycle. In the typical four-stroke engine, the intake and exhaust valves and the fuel injector are located in the cylinder head (see figure). Two valve systems are often used: two inlet valves and two outlet valves.

The use of the two-cycle cycle can eliminate the need for one or both valves in the engine design. The wash and intake air are generally supplied through the openings in the cylinder liner. The exhaust can be through pipes in the cylinder head or holes in the cylinder liner. The construction of the engine is simplified when an opening design is used instead of requiring exhaust valves.

WORKSHOP

A simple biodiesel processor

The simplest way to make a biodiesel processor is to use a 55 gallon (208 l) steel drum and some mixer. The blender may be a circulation pump, such as a suction device, or it may be an electric mixer for chemicals, specifically designed for mixing the drum. A pump or a blender will cost about $ 200 if you buy it again, but instead, you can make your own. With a bit of genius, you can build a biodiesel processor that is economical and efficient. Tim Garrett of Kelseyville, California, has created a processor of this type from mostly recycled pieces for less than $ 50. A simple biodiesel processor can be made from the following parts:

• The metal drum of 55 gallons (208l).

• 1/2 hp electric motor.

• Two pulleys that give approximately 250 to 400 rpm on the mixer.

• Belt driven by both pulleys.

• 2-inch (5 cm) wound rod mixer rod.

• One propeller made of two shelves welded to each side of a rolled rod of 2 inches. The shelf mounts look like two opposite "L" and form a spiral diameter of 14 inches (36 cm). Any metal in the form of a propeller would if it is made of steel with 12 or 14 gauges, • 3/4 inch (19 mm) ball valve for draining glycerine.

• A hinge and a piece of wood that acts as a belt tensioner.

• 2,000 W electric boiler.

• Boiler thermostat.

• Wood, screws, screws and other mounting accessories.

Attention note

These are dangerous chemicals when it comes to biodiesel. Methanol and bases are strong bases. They can endanger nerve endings and cause permanent damage. For this reason, chemical resistant gloves, apron, and goggles should be used when working with methanol and alkaline bases. Shoes, long-sleeved shirts, and long pants are needed.

Keep methanol and alkalis in clearly labeled containers. We recommend placing a skull and crossbones on them and writing something by "Danger! Poison Do not Eat!" With the content.

Sodium methoxide, the chemical combination of alkali and methanol, is even more toxic than the separate components. Keep these things away from exposed skin. Do not let kids play in or around equipment for biodiesel. Remember, although you produce two safe chemicals when producing biodiesel, use hazardous chemicals in the process.

Always think about safety when preparing a reaction to biodiesel. Have a faucet or a hose nearby. Keep vinegar on hand to neutralize methanol or spray paint. If you take the time to prepare and follow the safety instructions, your reaction to biodiesel will run smoothly and should not be a problem.

Fuel tax and engine specifications

If you live in the United States, you must pay a tax for any fuel on a road that is not subject to a tax on a pump. If you live outside the United States, it is advisable to consult the local tax authorities.

You are responsible for any damage that the engine could cause if you use fuel that does not meet the specifications of the manufacturer of your engine.The islands produce biodiesel from coconut oil, some countries experiment with biodiesel from cannabis oil, and many others use canola oil.

Millions of kilometers of road tests have been done with this fuel. Tests have shown less wear on the internal components of engines using biodiesel.

Biodiesel is a reliable and stimulating fuel that you can use.

If your diesel engine is of concern to you, you can install an additional fuel filter system from Racor or a similar manufacturer. After transporting more than 40,000 km of biodiesel from a used edible oil, we continue to choose and recommend biodiesel instead of toxic and carcinogenic diesel oil.

Appleseed 2.0 Biodiesel Processor Plans

The Classic Appleseed Reactor is an easy-to-assemble processor made from locally sourced materials that do not require welding. Maria "Girl Mark" Alovert is popular in its dynamic national workshops. It took almost four years to refine the original and design this style. B100 Supply uses the same program for its popular biodiesel processor kit.

In this version, the only tricky part to find locally is a bimetallic thermometer, and it is available on the Internet. The pipe is a black pipe, but the galvanized works if it cannot find the black tube. Brass ball valves are good, make sure the ball is metal inside. Some of the cheapest are metal and plastic bullets. Although the used water heater may work, it is much easier to start with a new one. You can spend more than a day cleaning one, and the material may not want to leave. If you look at it, you may find scratches and teeth or an inverted boiler that will allow your store to sell at a reduced price. You want to obtain the cheapest electric boiler available in the desired size. They are more expensive and have electronic controls, which would complicate the wiring of the boiler. For processors that consist of 40-gallon or fewer water heaters, I prefer low-fat or low-fat water heaters because it may be preferable to mix them. However, large water heaters are more comfortable to drain the glycerin with less mixed bio/glycerin.

When building your Appleseed biodiesel processor, remove the dip tube in the cold water inlet and the anode rod. The immersion tube is a plastic tube located under the nipple at the cold water inlet. Once the nipple is out, the dip tube is released by inserting a screwdriver into the plunger tube and rotating the handle from side to side. Once released, merely put the finger on the dip tube and remove it. The anode bar is usually located on the top of the heater under a plate and is generally labeled. So far, all I've seen is a 1 1/16 "hex. It's the same size as found on a Ford truck tire tool.

Biodiesel Processor Bill of Materials			
Item	Description	QTY	Notes
A	3/4" X 4" pipe nipple	3	
B	3/4" x 45 deg Elbow	1	
C	3/4" x close nipple	8	
D	3/4" x 3/4" x 1/2" TEE	2	Note 2
E	1/2" hose barb	3	
F	1/2" ball valve	2	
G	1/2" close nipple	3	
H	1/2" street elbow	1	
I	1/2" swing check valve	1	
J	3/4" x 1/2" bushing	1	
K	3/4" cross	1	
L	3/4" ball valve	4	
M	3/4" hose barb	5	
N	3/4" union	1	
O	1" x 3/4" bushing	1	
P	clear water pump	1	Note 1
Q	1" close nipple	1	

Biodiesel Processor Bill of Materials			
Item	Description	QTY	Notes
R	1" x 3/4" x 3/4" TEE	1	Note 2
S	3/4" braid reinforced PVC tubing	6ft	note 4
T	12 gauge power cord	6ft	
U	thermometer	1	note 3
V	3/4" x 1/2" x 3/4" TEE	1	Note 2
W	1/2" ID clear PVC tubing	10 ft	note 4
X	3/4" pipe plug	1	
Y	1/2" elbow	1	
Z	3/4" x 16" pipe nipple	1	
AA	3/4" elbow	2	
AB	1/2" x 12" pipe nipple	1	
AC	3/4" x 12" pipe nipple	1	
AD	30psi relief valve	1	note 5

Install the vent and relief valve at the top, away from your work area. During normal operation, the vent will emit methanol vapors, and you want them to go where they can dissipate without the risk of poisoning people or animals. The original NPT can be left or replaced with a pipe plug. It should be noted that when the temperature rises to about 20 ° C (200 ° F), the valve opens and drips; it must look like the vents in your work area. It is also necessary to disable the upper heating element.

The traditional Appleseed processor is the embodiment of KISS in the design of the DIY biodiesel processor. It was stupidly simple to build, and all the pieces come from your average hardware store. Well, this is the third major update of the design and is still easy to create. Only this time, we turned to experts in chemical engineering, hazardous materials management, fire safety, and electrical codes to make Appleseed safer and more efficient. It is still easy to build and is now the easiest DIY biodiesel processor to use.

Appleseed 3.1 has all steel lines to enhance safety. Its triangular configuration dramatically facilitates the installation of all steel pipes. It has a high-speed nozzle installed in the "A" pipeline that provides an

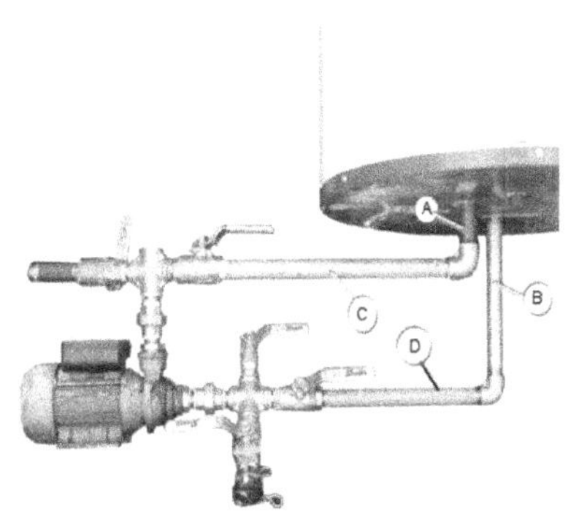

improved mix that reduces the amount of methanol we need and reduces the processing time. The combination has been so advanced that we can pump our methoxide as fast as possible without skipping hoops as in previous versions of Appleseed.

Pipe "A" is a special 3/4" x 4" schedule 80 pipe nipple with 1/2" npt female treads tapped in one end.

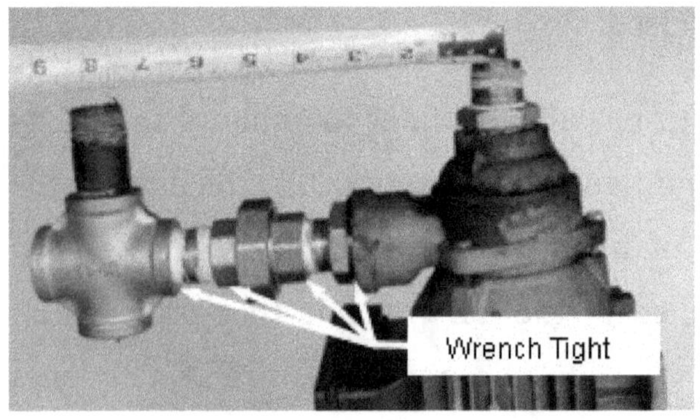

This allows us to screw our nozzle into the nipple, then screw the nipple into the reservoir, thus providing a nozzle almost flush with the bottom of the tank. They are available on the Internet, or you can request the nipple 80 from your local plumber and ask them to touch the nets it contains.

Hose "B" must be made special in the hardware store. This is a 3/4 "threaded pipe at both ends. To determine the length, measure the distance between the pipe axis C and the axis of the D pipe and add 4" for the length of the pipe B. Take simply the range. At the hardware store and they will make a pipe to fit you.

Once you have the custom pipe, install it (pipe B) and pipe A on the two ports furthest away from each other.

Tighten one down as normal, then tighten the other so that the difference in height from the water heater is the same as your original measurement.

Since we are using a triangular configuration we can use stock pipe nipples for C and D. They just need to be within 4 inches of the right size. We can then rotate the triangle align everything.

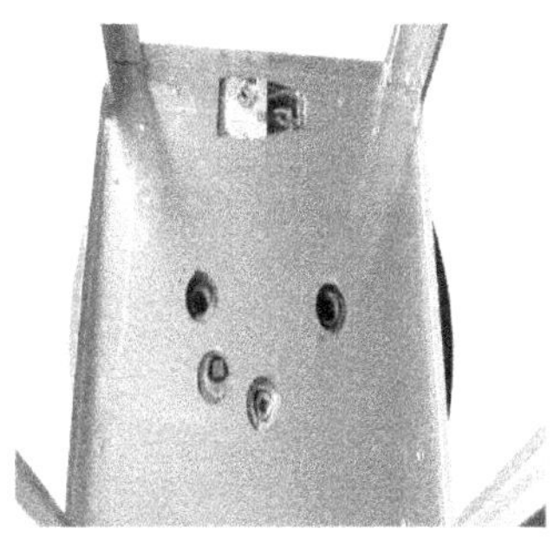

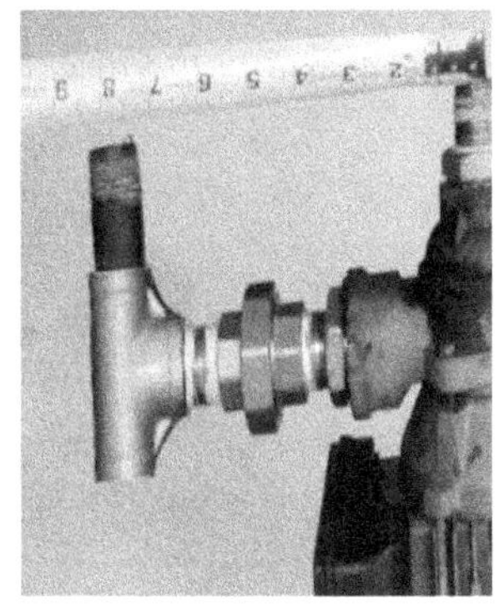

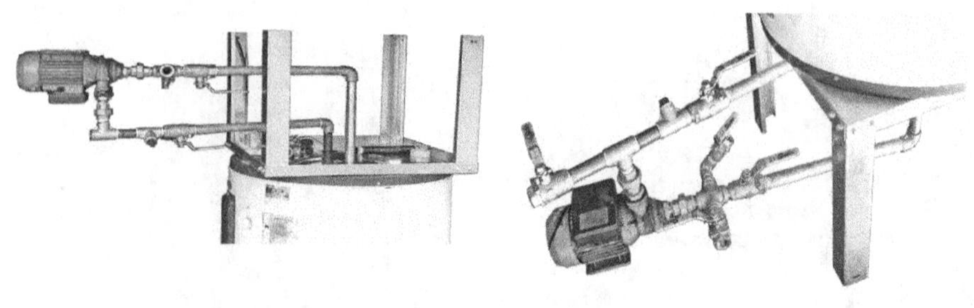

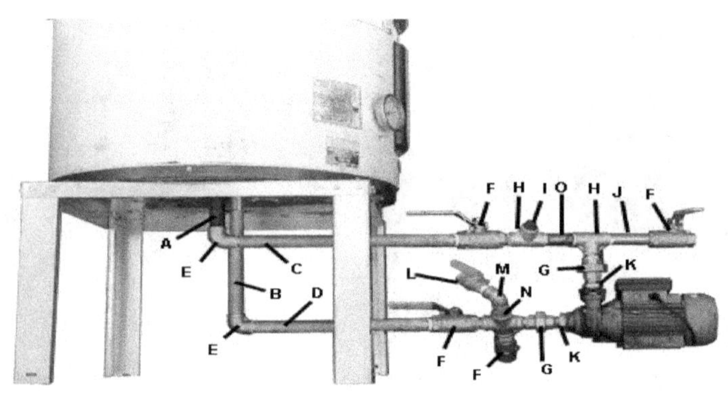

Bill of Materials

A - 1ea - Adapter Pipe

B - 1ea - 12" x 3/4" nipple **Note 1**

C - 1ea - 16" x 3/4" nipple

D - 1ea - 14" x 3/4" nipple

E - 2ea - 3/4" elbows

F - 4ea - 3/4" ball valves

G - 2ea - 3/4" pipe union

H - 2ea - 3/4" TEE

I - 4ea - 3/4" pipe plug **Note 2**

J - 1ea - 6" x 3/4" nipple

K - 2ea - 1" x 3/4" bushing

L - 1ea - 1/2" ball valve **Note 3**

M - 1ea - 1/2" elbow

N - 2ea - 3/4" x 1/2" bushing

O - 1ea - 3" x 3/4" nipple

    7ea 3/4" close nipple

Note 1 The size of pipe "B" varies depending on the pipe fittings you use and the specific pump you use. In my case it is very close to 12" long. Using two 6" nipples joined with a coupling allows you to adjust the length more finely

Note 2 This port is for future expansion. The other three pipe plugs are used to plug the unused holes in the water heater.

Note 3 The 1/2" ball valve is used for the sight tube.

Alternative design

If the construction is too difficult for you with your custom length pipe, an alternative design replaces the "D" pipe with a pex pipe and eliminates custom length pipes. PEX is resistant to biodiesel and methanol. It's resistant. If you use PEX "Shark Bite" accessories, you do not need special tools. However, these are disposable accessories, so be careful when using them. The use of pex inserts plastic pipes that can melt and empty the contents of the processors in case of fire. This is why all steel tubes are always preferred for plastic.

The nozzle is not optional. You need the nozzle, and the tube adapter or the processor will not work correctly. Fortunately, the nozzle can be extracted from the most competent plumbers. It is a programming nipple 80 with female thread pivoted at one end.

MAKING BIODIESEL AT HOME

Making biodiesel at home

Biodiesel is mainly derived from a chemical process called transesterification. This involves replacing the glycerol component of the oil used with alcohol, which is carried out in the presence of a catalyst. Although it sounds complicated, making biodiesel at home is a relatively simple process. Farms around the world have been using it for generations. It only becomes popular in the mainstream because of the growing need to find a viable alternative to fossil fuels. By using easily accessible ingredients, simple tools, and simple techniques, this can be done at home and take advantage of reduced production costs.

Method 1

This technique is useful for those who want to produce biodiesel at home in large quantities. This is a little complicated and requires us to take many precautions.

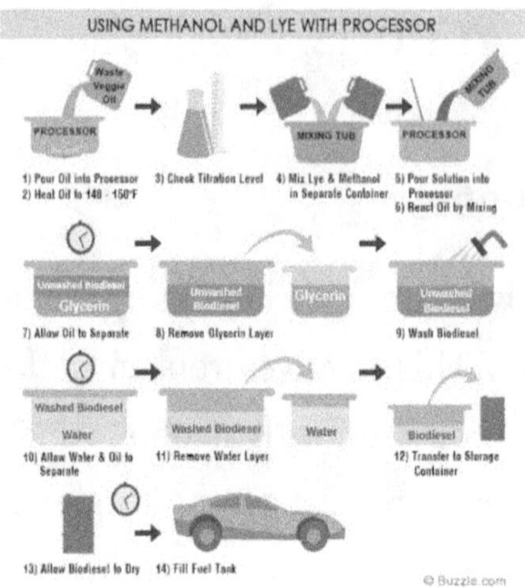

Things Required

- A large bucket

- A big bag for filtering the oil

- A catalyst like sodium hydroxide or potassium hydroxide (the latter is better)

- Methanol, waste, or new vegetable oil

- A processor for heating and mixing

- Isopropyl alcohol

- Measuring cylinders

- Titration apparatus (burette, pipette, conical flask, beaker, etc.)

- Phenolphthalein indicator

- Distilled water (DW)

- Respirators that are chemically resistant

- Long-sleeved gloves

- Eye protection like safety goggles, etc.

The procedure

1. First, you need to filter the oil before you start any complicated process. For this, it's an easy way to hook a porous pock to the pillar and hold a large bucket underneath. Oil is poured into these bags and filtered to accumulate in the trash below.

2. After collecting a large amount of oil in the trash (about 20 liters), take something from the glass as a titration sample, which is the next step in this process.

3. Since the oil is used in large quantities, it is essential to determine the amount of catalyst added. A significant deviation from the actual amount may cause the decomposition of the reaction. For this, a sample of the oil used to measure its acidity or the amount of fatty acids present in reaction with sodium hydroxide is tested.

4. Mix 1 ml of the liquid with 1000 ml of DW. 0.1 mol of NaOH is obtained. Fill this tray in the burette and remove the air pockets.

5. Now mix about 5 ml of vegetable oil with about 25 ml of methanol in a conical flask with pipettes and add a few drops of phenolphthalein to this mixture. Titrate it against the NaOH solution in the burette and continue turning until the mixture in the flask turns pale pink. At this point, the acid in the oil is neutralized, and the reaction is complete.

6. Record the reading of the amount of alkali used and find the required amount of this compound using the following formula:

Number of Moles (N) = Volume (V) x Molarity (M)

For example, if neutralizing 25 ml oil with methanol requires about 5 ml of lye (0.1 molarity), then the concentration of the latter would be:

No. of moles (NaOH) = 25 x 0.1 = 0.25 moles

This provides us with a rough idea of the number of chemicals required for making the biodiesel.

7. After that, the filtered oil is taken and added to the biodiesel processor. With this method, you can use a large amount of oil (about 5 gallons) to prepare an equivalent amount of biofuel.

8. Heat to about 150 ° F and allow cooling for a while. In a separate treatment tank, the methanol and alkali solutions are mixed to form a catalyst.

9. Using the appropriate compounds, this mixture is added to the vegetable oil, leading to the completion of a reaction that lasts at least a few days. Meanwhile,

periodically turn the tank closed to mix the oil with the catalytic converter adequately.

10. When the addition is complete, leave the container uncovered for several days. A thick layer of glycerine will appear on the bottom; a dark brown color will appear above, biodiesel.

11. This fuel is then extracted by careful separation using suction pumps and other methods. It is still dark and cloudy and therefore requires some washing with distilled water.

12. Place in a rotating container and add the water at a ratio of 500 ml of water to 1 liter of biofuel. Turn the container very gently for about a minute and stop the process. The water will be separated from the oil, removing impurities, and it will appear brown and muddy.

13. Collect the filtered oil and repeat the procedure until you have a transparent layer of water. It is the quality of the biodiesel to obtain, usable for your vehicle.

Method 2

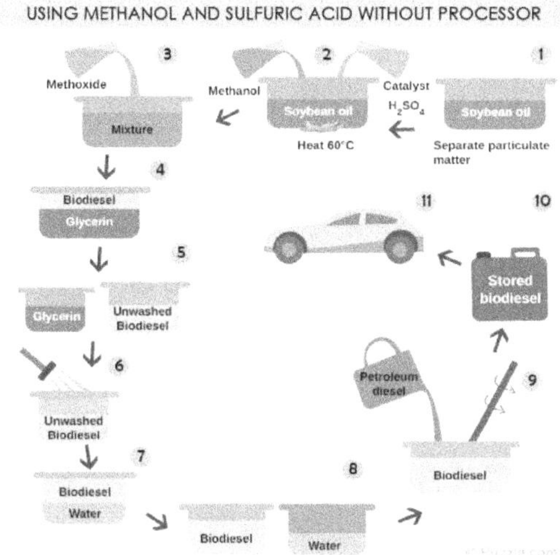

Things Required

- Vegetable oil (preferably soybean oil)

- A vessel for making the fuel

- A settling tank

- Filtering system

- 95% pure sulfuric acid

- 99% pure methanol

- A prepared mixture of methoxide

- Measuring cylinders, beakers, and pipettes.

The procedure

1. Filter the oil to remove any particles, such as fried food pieces. Use some filter screens. If you want to avoid this step of the process, you can buy fresh oil.

2. Heat the oil at about 60 ° C for about 15 minutes to remove any water.

3. Then place the oil in the storage tank and let it sit for 24 hours to allow separation. Or drop the water up or down.

4. Next, the oil should be accurately measured and warmed until all solids are melted. It is essential to gauge how the other ingredients to be added correctly would also be in the proper proportions.

5. Then, using a ratio of 8% of the total amount of oil, add at least 99% pure methanol. It is good if the purity of methanol is high.

6. Continue mixing the methanol in the oil for about five minutes. At this stage of the process, the mixture of oil and methanol will be unclear.

7. Next, for each liter of oil, add 1 ml of 95% sulfuric acid. Remember to be very careful taking all possible precautions when handling sulfuric acid as this can be extremely dangerous. Heat this mixture to 35 ° C and continue stirring.

8. Then remove it from the heat source and continue to mix gently for 2 hours. Let him stay for about 8 hours. If you notice that some of the mixtures have hardened at rest, warm up slowly.

9. Put half of the 12% volume of methoxide mixture into it, stir for 5 minutes, and let the oil settle down. The cloudy layer will start becoming clear, and this is the biodiesel portion of the mixture that will separate from the glycerin. The latter forms a darker part at the top, while the biofuel forms the light-colored brownish layer at the top.

10. The next step is called washing, wherein this fuel is filtered many times with water and other filtering equipment so that the impurities are removed, resulting in clean biodiesel. Then, it can be blended with petroleum diesel too in appropriate proportions, and finally used for the vehicle.

Method 3

USING METHANOL AND LYE WITHOUT PROCESSOR
1
Methanol
Lye
2
Stir the mixture
3
Vegetable oil
Heat 150°F
4
Heated oil
5
Methanol/Lye mixture
6
Shake
7
Unwashed Biodiesel
Glycerin
8
Glycerin
Unwashed Biodiesel
9
Water
Unwashed Biodiesel
10
Washed Biodiesel Water
11
Biodiesel
12
Stored biodiesel
13
Use in vehicle
© Buzzle.com

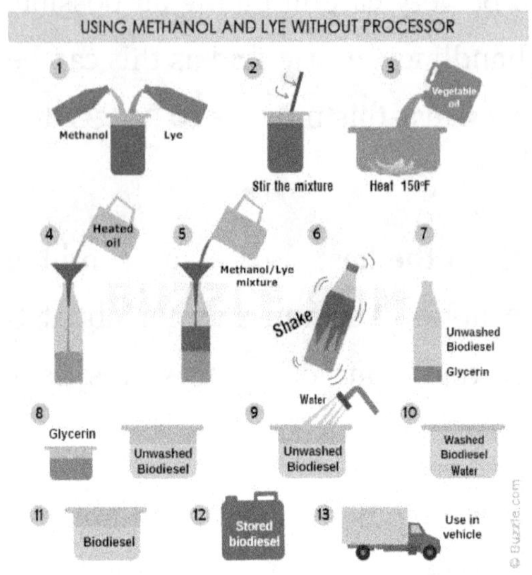

Things Required

This is a straightforward procedure to make biodiesel and requires readily available equipment.

- A bottle of vegetable oil (used or new)

- Sodium hydroxide or lye

- Methanol

- Funnel

- Stove or any heating equipment

- Glass container

- Non-aluminum lid

- Safety goggles or eyewear

- Hand gloves with long sleeves

- Digital scale to accurately measure the weights

- Respirator

Make sure that the place where you are undertaking this procedure has a good source of ventilation.

The Procedure

1. First, take about 300 ml of methanol in a glass container. Add about two teaspoons: sodium hydroxide or base in this alcohol in the container.

2. You must now make sure that the solution of the base dissolves completely in this mixture; For this activity, use an agitator to vortex the mixture. Attach the non-aluminum cap to the container and hold this mixture to one side — an alternative to the mixture of methanol and sodium hydroxide in the mixer.

3. Now take the vegetable oil hopper into a metal container and heat it to about 150 ° F. Let it cool slightly and pour it into a 3-liter clear bottle.

4. Add a mixture of sodium hydroxide and methanol to this oil in the flask using a funnel. Be careful when doing this step as the contents of the mixture are extremely harmful if it comes in contact with the skin and eyes or even when inhaled.

5. After adding the solution, close the bottle well and shake it for about a minute. Keep the bottle on a flat table and watch the changing density of the vegetable oil. At the bottom, a darker layer of glycerol will begin to form, separating from the lighter layer of vegetable oil to diesel oil. This is due to the substitution reaction, whereby the glycerol molecule in the oil is replaced by a fragment of alcohol to form the desired fuel. At this point, the diesel will be muted or mutilated.

6. Keep the bottle uninterrupted for several days; the fuel will start to appear more evident as the layer of glycerine finally settles.

7. Biodiesel can be removed by funnel or pipetted.

Advice, safety, and precautions.

• Try this procedure in small amounts, about 1 liter. By doing so, you will know what to look for when handling large quantities.

• The chemicals used in this process are dangerous. Therefore, you should take all precautions before doing it at home.

• The alkaline catalyst used (NaOH or KOH) is corrosive and can cause extreme irritation if it comes in contact with the skin and eyes.

• Methanol can cause immediate blindness and, therefore, is a hazardous type of alcohol. It can be easily absorbed by the skin and can cause death in extreme circumstances.

• When the aggregate catalyst reacts with a mixture of oil and methanol, sodium methoxide and gas are produced. The vapor of this gas is very toxic and acts as a means to paralyze the nerves if they breathe. Therefore, be sure to use a respirator that can block unwanted vapors and volatile substances.

Although the production of biodiesel in the home is not a difficult task, it is necessary to be patient to produce useful quantities of this fuel. Things can go wrong at any time, so it is unlikely that you will get the perfect biofuel quality. If you follow all the steps correctly and take all the safety measures, it will be effortless for you to produce large quantities of this type of fuel, which will ultimately help our planet Earth to preserve the environment

Making Biodiesel at Home 2

Before entering the details of producing small quantities of biodiesel in the house, we must first emphasize the importance of security. Chemicals used in the biodiesel process are dangerous, and if not appropriately handled, appropriate safety measures can cause severe injury or death. Be careful and make sure you are in a well-ventilated area with access to running water.

Required materials

• 1 liter of vegetable oil - new (SVO) or used (WVO)

• NaOH (alkaline / caustic soda), at least 6 g. This is often used as a drain cleaner and can often be found in a local store.

• Methanol (at least 250 ml). It used as an antifreeze; it can often be found in the car supply stores.

• Plastic bottle from 1 to 2 liters: I often use an empty vessel for vegetable oil

• Measuring container

• Scale

Or

• A teaspoon

• Storage tank for methanol and NaOH (methoxide). Not plastic. Glass is recommended for difficult conditions of use.

• Funnel

Heating your WVO

If waste oil is used (WVO), it runs liters and heats up to at least 120 cubic degrees to remove all water. If there is water, the oil spit, and work, when it is separated, it will stop. Be careful; this can be a very violent process. Then let it cool down

If new vegetable oil is used, it must not contain water so hot simply 55d when it is ready.

Making the methoxide

WARNING making methoxide is dangerous. Methoxide is highly toxic. Therefore, the safety equipment Design and working space must be carefully considered before use and should wear protective clothing and respirator during handling. The amount intended to be used immediately needs to be created only.

Take 250 ml of methanol and add 4 g (about half a teaspoon) of NaOH.

If waste vegetable oil (WVO) is used, only 6g-7g NaOH is applied (approximately 1 tsp)

Methanol and NaOH are not easy to mix. Start with methanol at body temperature (hot). Keep in mind that by combining two chemicals, the temperature will increase. Do not be afraid it's normal. You need to ensure that all NaOH dissolved in methanol, this could take more than ten minutes.

After all OF NaOH dissolved, you may need to top up with fresh methanol because the process can cause some evaporation.

Making the biodiesel

When the oil temperature falls to 60 degrees Celsius or less, pour a liter of oil into your dry plastic container with a funnel. Take methanol / NaOH (methoxide) and add to the oil. Make sure the container is well closed and shake vigorously for about 15 seconds.

Leave your biodiesel "set." You will notice that after 10 minutes glycerine or "soap" will be removed from your mixture. It will take a day or two to separate the biodiesel. You will see two defined layers: biodiesel and glycerine. Typically, the glycerine layer is about the same or slightly higher than the amount of methanol used.

Now remove the biodiesel from the glycerine container and be ready to wash.

BIODIESEL WASHING

Although glycerine or "soap" and water are separate from your biodiesel, you will still need to wash. Do not shake the unwanted biodiesel forcefully, as it will create an emulsion that can take days or even weeks to completely separate. Soft focus is what is needed.

Wash one:

- Pour 1 liter of biodiesel into a bottle of a clean, dry planer.
- Pour gently into 500 ml of water (body temperature).
- Replace the bottle cap.
- Now, gently turn the end of the bottle to the end for about 30 seconds.
- After 30 seconds, place the bottle upright.
- Only if you have been mild, water and biodiesel will be immediately separated.
- You will notice that the water is not clear.
- Remove the top and use your thumb as a cap, flip the bottle and drain the water using your thumb as a valve.
- You have finished washing one.

Wash two:

- Pour another 500 ml of water and repeat washing, except that it turns gently for about 1 minute.
- Drains as in wash one.
- You have finished washing 2.

Wash Three:

- Again, pour another 500 ml of water and gently shake the bottle for about a minute.

- When water and biodiesel are separated, discard the water in the same way as before.

Wash four:

- 500 ml of water and a little more stirring for about 1 minute.

- After separating the water and the biodiesel, drain as indicated above.

Wash five

If the washing has been done successfully, the water should be almost transparent. Keep in mind that during your subsequent wash, you should be able to shake it violently, even if it will take much longer to separate it because the water forms small insects in the biodiesel that take time to calm down.

Your washed biodiesel will be VERY CLOUDY and much lighter in color than unwashed biodiesel. After a day or two of sedimentation and drying, it will disappear.

Drying your Biodiesel

You must remove all that water from your biodiesel before using it in a diesel engine or risk damaging the engine. The oldest method of drying is settling.

In this method, the water settles to the bottom of the tank or container over time and can be sucked out using a small pump or siphon. For small batches, it can take up to a day for the water and biodiesel to completely separate.

Over time the water will evaporate out of the biodiesel; however, if biodiesel is let in a moist or wet environment, this may not be suitable.

Once the water has all be removed you Biodiesel is now ready for use! Enjoy!

METHANOL RECOVERY

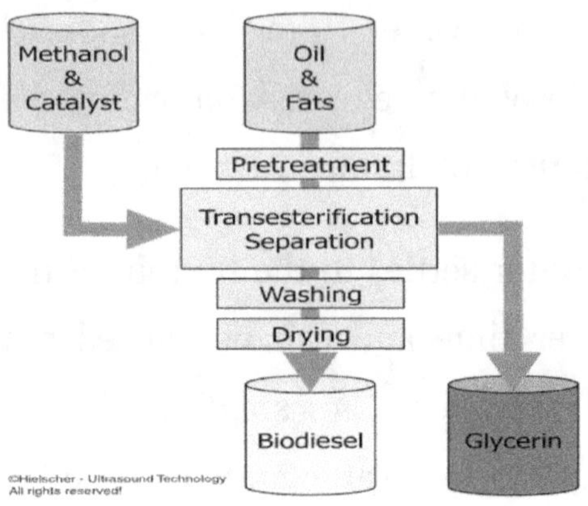

Depending on the type of oil used, it takes 110 to 160 ml of methanol to produce 1 liter of biodiesel.Excess methanol (20% or more of logical volume) is required to complete the conversion process. Most of the excess methanol is in the glycerol phase of the biodiesel reaction and can be recovered merely by boiling it in a closed container with an outlet leading to a simple condenser. Methanol boils at 65 ° C but generally starts to evaporate well before boiling point. To coat the methanol with glycerol, heat the crude glycerol to between 65 ° C and 70 ° C (149-158 ° F). As the methanol evaporates, leaving an even smaller proportion of methanol in the mixture, the boiling point will increase; therefore, it is necessary to continue raising the temperature to maintain the vaporization of methanol. The liquid begins to foam as soon as the temperature reaches 100 oC (212 oF); stop heating, or you will get a sparkling brown by-product in the methanol condensate. Most of the methanol should have been recovered at this time. Be sure to use a sealed container for all efforts to recover methanol. Methanol vapors are toxic and pose significant health risks if inhaled.

REACTORS

For biodiesel production, three general types of reactors are used: batch reactors, semi-continuous reactors, biodiesel ultrasonic reactors, and continuous flow reactors.

Series reactors

The batch reactor may be a container equipped with some agitation. The tank is filled with reagents for the process (in this case the oil, the alcohol, and the catalyst), then the mixer runs for a while. Once the required time has elapsed, the contents of the reactor are emptied and processed.

Continuous flow reactors

The most common system of constant flow in biodiesel production is the continuous mixing reactor (CSTR). As shown in Figure 1, CSTR seems, at first glance, identical to a series reactor. The actual reactor can often be the same, but additional controls are needed for the reactor configuration in the continuous flow system. Some continuous flow installations may operate in serial or constant mode.

The reagents and the product (a mixture of different chemicals, including unreacted reagents) are continuously extracted into the CSTR. Proper mixing is necessary to ensure uniform chemical composition and temperature. The continuous flow process typically requires complex process controls and online monitoring of product quality.

Biodiesel ultrasonic reactors.

Ultrasound is a useful tool for mixing unloaded fluids. When producing biodiesel, an appropriate mix is needed to create sufficient contact between vegetable oil/animal fat and alcohol, especially at the beginning of the reaction. The ultrasonic waves cause intense mixing so that the reaction can occur much more quickly.

Ultrasound transmits energy to the fluid and creates strong vibrations that form cavitation bubbles. When the bubbles burst, the liquid suddenly contracts and the ingredients mix in the bubble area. This high energetic activity in the fluid can significantly increase the reactivity of the reaction mixture and shorten the reaction time without high temperatures. This reaction can be obtained at room temperature or slightly above. Since it is not necessary to heat the mixture, it can save energy.

SECURITY MEASURES

Exposure and safety of chemicals

Methanol (flammable and toxic alcohol) and laundry (acid and caustic bases) are two dangerous chemicals needed to turn vegetable oil into biodiesel. Excessive exposure to methanol can cause neurological damage and other health problems. Methanol is also a severe fire hazard.

Treatment of by-products.

Biodiesel processors generate significant amounts of crude glycerol byproducts (approximately one gallon of waste containing glycerol over five gallons of biodiesel produced). Glycerol and wastewater should be handled and disposed off.

Liability in case of vehicle/equipment failure

When a small manufacturer pours the first gallon of domestic fuel into the tank, this manufacturer assumes responsibility for the future performance of the equipment concerned. Thousands of small manufacturers around the world successfully drive diesel equipment with domestic fuel, but users also need to understand that problems can arise.

Handling of chemicals

Methanol

Methanol is toxic and should be handled and used in a well-ventilated area. Inhalation or ingestion of methanol may be very harmful at higher concentrations and may result in death or blindness. This is particularly harmful to the eyes. When handling the product, wear chemical-resistant goggles, clothing, and gloves. If the concentration in air exceeds 200 ppm, a respirator is required, preferably with a face mask. (Vessel respirators are not useful for regular use in methanol pairs).

Sodium hydroxide and potassium hydroxide

Sodium hydroxide (NaOH) and potassium hydroxide (KOH) are corrosive and can be fatal if ingested. (These chemicals are called "alkaline" or "catalytic"). Contact with the skin can cause severe burns and rinse the affected area well with water or a diluted vinegar solution. Inhalation of solid NaOH or KOH is possible if the material is reduced to dust-like particles. Each of these situations is critical and requires immediate medical attention. Sodium hydroxide and potassium hydroxide must be separated from the water because the water will inhibit the biodiesel reaction and cause the release of heat due to mixing, which could cause a fire in the adjacent materials. NaOH, KOH and concentrated solutions should never come in contact with aluminum because they will generate explosive hydrogen gas.

Appropriate safety equipment for working with NaOH or KOH includes bent gloves, dust goggles, dust mask or respirator, long pants, and shoes. Eye irrigation and emergency showering are also recommended within 25 meters of the work area. (The eyewash station can be as simple as a garden hose or a dedicated garden tap that moves fresh, regular water to the eye.) A manual design allows the affected person to use their hands to keep their eyes open while blushing

General Fire Safety

No flames, smoking or sparks anywhere near the production area.

Safety Gear Summary

The following safety gear should be on hand each time you brew biodiesel:

• Chemical-resistant gloves (butyl rubber is best for methanol and lye)

• Chemistry goggles (indirect vented)

• Face shield

• Dust mask or cartridge respirator

• Plumbed eyewash station

• Small spray bottle with vinegar for neutralizing lye spills

• Access to running water

• Telephone in case of emergency and emergency telephone numbers

• Fire extinguishers (20-lb ABC)

• Absorbent material and spill-containment supplies

REFERENCES

CHAPTER 1

- Small Scale Biodiesel Production - Wilson Collegehttps://wilson.edu/sites/default/files/uploaded/completedmanualbiodieselfinaledit1.pdf ,February 17, 2019

- "Procurement of Motor-Pumps for Chad [Tender Documents: T35191283]." 2016. MENA Report, N/A.

- 7 Agriculture Adjutants Biodiesel is used as a carried forhttps://www.coursehero.com/file/pdfare/7-Agriculture-Adjutants-Biodiesel-is-used-as-a-carried-for-pesticides-and/February 17, 2019

- The Facts on Sustainability - Biodiesel Sustainability Bloghttps://www.biodieselsustainability.com/better-for-the-environment/the-facts-on-sustainability/February 17, 2019

- Can I make my own biodiesel at home and is it safe?http://www.makebiofuel.co.uk/can-i-make-my-own-biodiesel-at-home-and-is-it-safe/February 17, 2019

- Top 15 Unexpected Uses For Biodiesel | Gas 2https://gas2.org/2008/03/26/top-15-

unexpected-uses-for-biodiesel/February 17, 2019

CHAPTER 2

- Biodiesel Production Principles And Processes – Biodieselhttp://www.biodieselproject.com/biodiesel-news/biodiesel-production-principles-and-processes-2.htmlFebruary 17, 2019
- Biodiesel Production Principles and Processes – eXtensionhttps://articles.extension.org/pages/27137/biodiesel-production-principles-and-processes February 17, 2019
- Introduction Biodiesel - Biodiesel Solutions – Biodiesel … http://www.biodieselproject.com/biodiesel-news/introduction-biodiesel.htmlFebruary 17, 2019
- Biodiesel as an alternative fuel for compression ignition … https://www.irjet.net/archives/V4/i9/IRJET-V4I9267.pdfFebruary 17, 2019
- Chapter 2 Introduction to Biodiesel Production - MSU Extensionhttp://msue.anr.msu.edu/uploads/files/biodiesel_production.pdfFebruary 17, 2019
- Chapter 2 Introduction to Biodiesel Production -

- canr.msu.eduhttps://www.canr.msu.edu/uploads/files/biodiesel_production.pdfFebruary 17, 2019
- Oil may be obtained from the fruit pulp and pit It has ahttps://www.coursehero.com/file/p3k8g3e/Oil-may-be-obtained-from-the-fruit-pulp-and-pit-It-has-a-high-nutritional-value/February 17, 2019
- Chicken House Water Pumps - TSI Poultryhttp://tsipoultry.com/poultry-farm-supplies/chicken-house-water-pumps/February 17, 2019
- http://www.clayton.edu/portals/690/chemistry/inventory/MSDS%20phenolphthalein.pdfFebruary 17, 2019

CHAPTER 3

The Chemistry of Biodiesel | Biodiesel Project | Goshen https://www.goshen.edu/academics/chemistry/biodiesel/chemistry-of/February 17, 2019

NEFA (Non-Esterified Fatty Acid) - pacbio.comhttps://pacbio.com/biomarker/assay-detail/368/February 17, 2019

Basic Chemistry of Biodiesel - attra.ncat.orghttp://attra.ncat.org/attra-pub/download.php?id=321February 17, 2019

CHAPTER 4

CHAPTER 1 INTRODUCTION 1.1 INTRODUCTIONhttp://umpir.ump.edu.my/id/eprint/8436/8/A%20study%20of%20the%20spray%20characteristic%20for%20valve%20covered%20orifice%20diesel%20nozzle%20injector%20using%20CFD%20-%20Chapter%201.pdfFebruary 17, 2019

diesel engine | Definition, Development, Types, & Factshttps://www.britannica.com/technology/diesel-engineFebruary 17, 2019

Experimental Analysis Of Single Cylinder Diesel Enginehttps://www.ijert.org/phocadownload/V2I7/IJERTV2IS70284.pdfFebruary 17, 2019

Diesel engine – Wikipediahttps://en.wikipedia.org/wiki/Diesel_engineFebruary 17, 2019

CHAPTER 5

GoPower - the-eye.euhttps://the-eye.eu/public/Strategic%20Intelligence%20Network/engineering/energy/Making_Biodiesel_%28Tickelll%29.pdfFebruary 17, 2019

eBook - How to Make Bio-Diesel - [PDF Document]https://vdocuments.site/ebook-how-to-make-bio-diesel.htmlFebruary 17, 2019

How to Make Biodiesel Fuel-1 - DocShare.tipshttp://docshare.tips/how-to-make-biodiesel-fuel-1_585fb500b6d87f96988b6f49.htmlFebruary 17, 2019

Appleseed 2.0 Biodiesel Processor Plans | The Appleseedhttp://www.make-biodiesel.org/The-Appleseed-Biodiesel-Processor/appleseed-2-0-biodiesel-processor-plans.htmlFebruary 17, 2019

Appleseed 3.1 Biodiesel Processor | The Appleseed http://www.make-biodiesel.org/The-Appleseed-Biodiesel-Processor/appleseed-3-1-biodiesel-processor.htmlFebruary 17, 2019

CHAPTER 6

How to Make Biodiesel at Home – HelpSaveNaturehttps://helpsavenature.com/how-to-make-biodiesel-at-homeFebruary 17, 2019

Make Biodiesel at Home - Make Biofuelhttp://www.makebiofuel.co.uk/make-biodiesel-at-home/February 17, 2019

CHAPTER 7

Reactors for Biodiesel Production – eXtensionhttps://articles.extension.org/pages/26630/reactors-for-biodiesel-productionFebruary 17, 2019

Safety in Small-Scale Biodiesel Production –
eXtensionhttps://articles.extension.org/pages/29265
/safety-in-small-scale-biodiesel-productionFebruary
17, 2019

CHAPTER 8

Biodiesel Safety and Best Management Practices for
Smallhttps://www.canr.msu.edu/uploads/files/Fuels
/Biodiesel%20Safety%20BMPs%20for%20Small%20
Scale%20Use%20Production.pdfFebruary 17, 2019